START WITH A QUESTION

Are All Zebra Patterns the Same?

FAQs About Living Things

Cynthia O'Brien

A Crabtree Forest Book

Crabtree Publishing
crabtreebooks.com

Notes for Readers and Class Discussion

This book is designed to teach readers about core subject areas, build curiosity, and inspire further investigation. Readers are encouraged to build upon what they already know about the subject and engage in topics they want to learn more about. Here are a few guiding questions to prompt reflection and discussion. Possible answers appear in red.

Before Reading:

Read the title and look at the table of contents. What do you already know about living things?

- *I know that living things are classified into different groups with similar characteristics.*
- *I know that human activities affect plants and animals.*

What would you like to learn about living things?

- *I would like to learn about the variety of living things around the world.*
- *I would like to learn how living things depend on each other.*

During Reading:

Pause after reading each page or chapter. What questions do you have about what you read? What are you curious to know?

- *I wonder what scientists are doing to learn more about how trees in a forest communicate and help each other.*
- *I am curious to know how animals, like insects, adapt to Arctic winters.*

Make connections to what you are reading. How is this like something you already know?

- *I know that dinosaurs are extinct.*
- *I know that some animals evolve over time to better suit their environments.*

After Reading:

Recall key details about the book. What was the author trying to teach readers?

- *The author was trying to teach readers that scientists ask questions to learn more about living things.*
- *The author was trying to teach readers that living things are interconnected.*

How did the images and captions help you understand more?

- *The photographs helped me see the wide variety of living things around the world.*
- *The captions helped me understand more details about living things.*

Table of Contents

Think Like a Scientist

One of the best ways to find out about anything is to inquire, or ask questions. This is how scientists investigate all subjects, from living things on Earth to the planets in space. They inquire, conduct experiments, and draw conclusions. The investigation may stop there, but often other questions arise during the process. The universe is still full of mysteries to explore.

What are living things? They are familiar animals and plants, but also include **fungi**, such as mushrooms, and **protists**. Even bacteria—so tiny they can only be seen with a microscope—are a type of living thing. You can ask all kinds of questions to learn about living things. For example, do sea creatures communicate underwater? If so, how do they do it? Are there plants that eat meat? Can insects survive in the Arctic?

Have a look at the questions and answers that follow. You will discover some amazing facts about living things, and you may think of even more questions to ask. Scientists never stop being curious!

Protists include animal-like organisms, such as amoebas, that can be seen through a microscope.

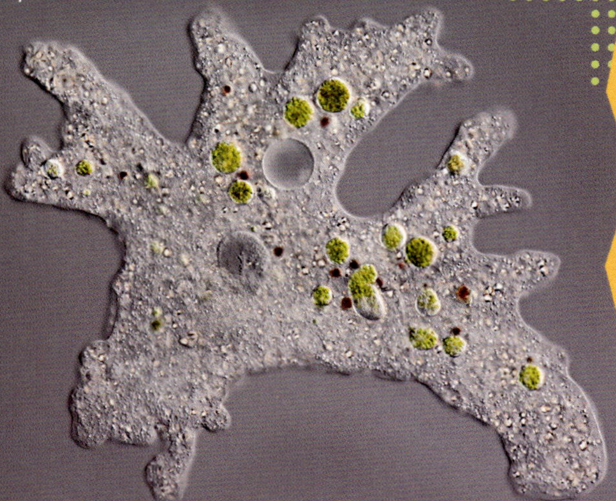

There are about 8.7 million types of plants and animals on Earth. Only about 1.2 million of these have been identified.

CHAPTER 1: QUESTIONS ABOUT MARINE ANIMALS

Can animals talk underwater?

Millions of creatures live in water all or some of the time. Some live in salty oceans and seas while others live in freshwater lakes, rivers, and streams. Scientists study life underwater using advanced technology, such as autonomous underwater vehicles (AUVs). These robots collect information to help scientists analyze how sea creatures behave, including how they communicate.

Fish communicate with each other in many ways, including body movements, electrical or chemical signals, and sounds. They communicate to attract mates, to identify each other, and to warn predators to stay away. They also warn each other about predators and help each other find food. Catfish, for example, release chemicals called pheromones to identify each other and send danger signals.

Many fish use color to send signals. Flatfish change their color to blend in with their surroundings, while sticklebacks use their bright colors to stand out and attract mates. Some minnows make red marks appear on their bodies during breeding season. Cuttlefish can change their color and shape to avoid predators.

Schools of fish use motions such as head shakes and twisting their bodies to communicate to each other as they swim in a large group.

Small, fish-like robots may help scientists to understand more about fish behavior.

Sound is a very useful way to communicate underwater because it travels quickly through water and over long distances. Fish have ears inside their heads and easily pick up sound signals. To make sounds, many fish use a muscle to vibrate their **swim bladder**. Others, like the seahorse, rub bones together. Clownfish make chirping and popping sounds by snapping their teeth.

Whales are vocal creatures and can be very social. Many species, such as dolphins and belugas, tend to **migrate** and hunt in groups called pods.

When whales communicate, they use a series of clicks and whistles. A few whales, including blue, fin, bowhead, humpback, and minke whales, produce sounds in patterns called "songs." Male humpbacks are the most famous for these songs and can produce them for many hours at a time.

Scientists do not know for sure why whales sing, but they have ideas. Whales may sing to attract a mate. They could be talking to their young or planning a migration. Whales might also sing to announce their presence to other whales and protect their territory.

Dolphins use unique whistles as a way to identify each other.

Humpback whales may breach, or break through the water's surface, to show off to other whales or even to create a powerful sound to communicate.

Can fish see in the darkest part of the ocean?

The ocean is brightest closest to the surface. Below 656 feet (200 m) deep, less and less sunlight gets through the water. At around 3,281 feet (1,000 m) deep, the ocean becomes completely dark. Deep-sea fish have eyes that have adapted to this environment.

Most are only sensitive to blue light, which travels the farthest in water. The blue light they see is made by deep-sea creatures using **bioluminescence**. This light helps deep-sea fish hunt, protect themselves, and communicate.

*Scientists discovered that the dragonfish has **evolved** to make and see red light. This allows these fish to communicate with each other and see red-colored prey at depths up to 5,000 feet (1,500 m).*

DID YOU KNOW?

There is life in the deepest part of the ocean. The Challenger Deep, in the western Pacific, is about 35,876 feet (10,935 m) deep. It is the deepest known part of the ocean. Sea cucumbers and other tiny organisms, such as worms, live there in darkness.

How long can a whale hold its breath underwater?

Whales must come to the surface to breathe air through blowholes on their heads. Some whales can spend long periods underwater before coming up for air. Sperm whales, for example, regularly swim underwater for about 90 minutes at a time while hunting. But this is less than half of the underwater record of the Cuvier's beaked whale. These whales can swim underwater for 222 minutes and dive as deep as 9,816 feet (2,992 m). To do this, whales store oxygen in their blood and muscles. They also conserve oxygen while hunting by slowing their heart rate, allowing their body temperature to drop, and shutting down organs, such as the liver, for short periods of time.

Beaked whales have sleek bodies that let them glide underwater, conserving energy.

Some tree frog species adapted to live almost entirely in trees. They absorb water and oxygen through their skin.

How can frogs live in the water and on land?

Frogs are amphibians—cold-blooded, scaleless vertebrates that can live in water and on land. A vertebrate is an animal with a backbone. Frogs have lungs to take in oxygen when they are on land. When they are in water, they absorb, or take in, oxygen through their moist skin.

Frogs lay their eggs in or near water, or in a moist place. These eggs develop into **tadpoles** with organs called gills to breathe underwater. As tadpoles grow into adult frogs, they develop legs and lungs, allowing them to adapt to life on land.

Amphibian is a word that comes from a Greek word meaning "double life."

Is it an alligator or a crocodile?

Alligators and crocodiles are amphibious reptiles—meaning they live on land and in water—and are both part of a large reptile group called crocodilians. But there are key differences. Crocodiles are usually more aggressive than alligators. Crocodiles have salt **glands** that allow them to live in salt water or fresh water, so they can be found in warmer parts of North and South America, Africa, Asia, and Australia. Alligators, on the other hand, are found only in the southern U.S. or the lower Yangtze River in China. Alligators are often gray or black, while crocodiles are olive green or brown. The jaws and teeth of the two reptiles also differ.

DID YOU KNOW?

Crocodilians are ancient! These reptiles have been around for about 230 million years. They are descendants of prehistoric reptiles like the ferocious *Deinosuchus* that preyed on dinosaurs. Other crocodilians include caimans and gharials.

Deinosuchus

An alligator has a wider, rounder snout. When its mouth is closed, only its top teeth are visible.

A crocodile's snout is pointier. Some bottom teeth are visible on a closed mouth.

Do animals ever help other species?

Mutualism happens when two species benefit, or are helped, from interacting with each other. Oxpecker birds, for example, perch on zebras and rhinos and eat the ticks and flies on their bodies. They also alert the animals with a loud warning call if there are predators nearby. In return, the birds get an easy and steady supply of food.

Oxpecker

During the course of a year, more than 1 million wildebeests go on a continuous migration in eastern Africa. Many thousands of zebras go too. The animals follow the rains to find fresh grasses to eat. They travel in a large loop that takes them hundreds of miles across Tanzania and Kenya. As they graze, the zebras eat the longer grass and leave the short grass for the wildebeests.

The animals also help each other during the migration by watching for predators such as lions and hyenas. Crossing the large Mara River is dangerous because massive Nile crocodiles lie in wait. The wildebeests rely on the zebras' keen eyesight, while the zebras depend on the wildebeests' excellent hearing and sense of smell.

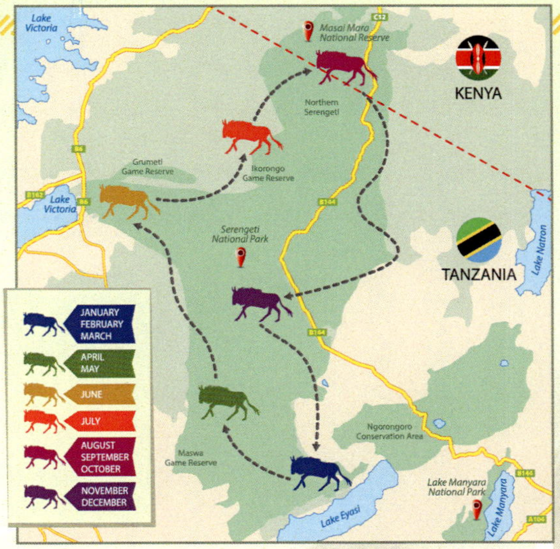

The wildebeest and zebra journey is often called the Great Migration because of the great numbers of animals that travel each year.

How do animals adapt to harsh habitats, such as the desert?

Animals have many ways to survive in the desert. Kangaroo rats live in dry and desert areas of North America. They can live without drinking water because they get the water they need from seeds. Camels can go a long time without water or food. They store fat in their humps that their bodies can use for energy. They also have long eyelashes to keep sand out of their eyes.

The kangaroo rat's kidneys take water from its urine, while its nose recaptures any water that might be lost when breathing.

Lava lizards are often seen close to large groups of marine iguanas. They eat the flies that land on the marine iguanas.

DID YOU KNOW?

Marine iguanas have amazing adaptations. The marine iguana is the only lizard that has adapted to spend some time in the ocean. It lives only in the Galapagos Islands and feeds on algae **that grows in shallow water. It has a flattened tail to help it swim and salt glands that enable it to remove the salt from the water.**

Can rescued animals survive in the wild?

Sanctuaries in Thailand are returning rescued elephants to the wild, but the process can take at least two years. The elephants must be released in a group into a large, but enclosed area. There, they get used to the wild again and teach their young. Eventually, they can be released altogether.

Elephants are slowly introduced to other elephants in controlled environments to make sure the group gets along before being released.

The same slow approach works for rescued lions. The adult lions must have cubs and form a group called a pride to be able to survive in the wild. Lion cubs stay with their mothers for at least two years. Female cubs may stay with the pride their whole lives.

How do chameleons change color?

Most chameleons are brown or green. But they can change their skin to bright colors to attract mates, defend territory, or compete with each other. Chameleons do this using **nanocrystals** in their skin cells, which reflect light, as well as skin **pigments**. Chameleons change the distance between the nanocrystals, making them reflect different **wavelengths** of light.

When chameleons are relaxed, their nanocrystals are close together and reflect blue light. The blue light combines with yellow from the chameleon's skin pigment, making the skin appear green. To create bright colors, chameleons push the nanocrystals farther apart. This allows them to reflect longer light waves, such as red and yellow. Chameleons can quickly appear shades of blue, orange, yellow, and other colors.

Unlike popular belief, chameleons do not change their color to match any background or pattern. Their natural color already blends in.

Brighter colors are a sign of a chameleon's strength.

Are all zebra patterns the same?

Just like human fingerprints, zebras have unique patterns of white stripes. However, each zebra species—plains, mountain, and Grevy's—has its own stripe pattern. This may help zebras recognize other zebras of the same species.

It is not certain why zebras have stripes. It could be that they help keep the animals cool, creating different air currents as sunlight shines over white, which reflects, and black, which absorbs. A herd of zebras also presents a confusing blur of pattern for predators, such as lions.

A herd of zebras is also called a dazzle. Their stripes may dazzle, or confuse, insects from biting and predators from attacking.

When elephants recognize their friends, they greet each other by rubbing their bodies together and linking trunks.

Why do elephants have good memories?

Elephants can remember family members they have not seen for many years and even people they have previously met. The oldest female elephant teaches her herd members routes to watering holes, places to find food, and how to communicate—all from her memory. Elephants are also capable of emotions such as joy and grief.

Though an elephant's 11-pound (5 kg) brain is the largest of all land animals—with about three times as many **neurons** as the human brain—most of its neurons are in the cerebellum, the area that controls movement. The cerebral cortex, the part of the brain that controls memory, has fewer neurons. So scientists are still investigating why an elephant's memory is so good.

Do wild animals like to play?

Researchers are trying to discover if animals play for fun or for other reasons. For example, wolves often play fight. They may do this to learn how to defend themselves or to form social bonds. Bear cubs play with young trees, bending them and jumping on them. This seems to be just for fun. Monkeys bang rocks together and roll them around. This may be just for fun, or to learn how to use the rock like a tool to break open nuts.

DID YOU KNOW?

Animals have play signals! Many animals ask each other to play. Chimpanzees make a "play face" by smiling with just their bottom teeth. A young male elephant will put its trunk on another elephant's head to say he wants to play.

Elephants also seem to play for fun, chasing and splashing each other and using a muddy riverbank as a slide.

Are birds really dinosaurs?

In 1861, a German paleontologist discovered a **fossil** of a feathered creature, later named *Archaeopteryx*. For a long time, everyone believed this 150-million-year-old fossil was the first bird. *Archaeopteryx* had feathers and a wishbone, a forked bone that helps birds to fly. However, *Archaeopteryx* also had a bony tail and teeth like non-birdlike dinosaurs.

In the 1990s, a fossil hunter in China discovered another feathered **theropod** dinosaur, later named *Sinosauropteryx*. It began to be clear that many dinosaurs had feathers. In 2013, paleontologists found yet another birdlike fossil, *Aurornis*. This fossil was 10 million years older than *Archaeopteryx*.

These fossils are helping scientists to understand how small theropods evolved to become modern birds. This evolution likely began to happen in the late **Jurassic period**, about 165 million years ago. Dinosaurs had snouts and teeth, but gradually, birds lost teeth and developed beaks. Birds also developed wings and different kinds of feathers. *Confuciusornis*, a fossil found in China, lived more than 120 million years ago. It is an early example of a birdlike dinosaur with a toothless beak, wings, and feathers that probably enabled it to fly.

A fossil of a dinosaur species known as Yi qi, which means "strange wing," was found in 2007. These dinosaurs had batlike wings.

The Tyrannosaurus rex was a theropod, making it a distant relative of birds. But birds evolved from much smaller theropods.

The feathered therapod's forearms probably evolved into wings.

No one is really sure why birds survived when land dinosaurs did not. Most birdlike dinosaurs were small, needed less food, and reproduced quickly. Birds also eat many different foods, which would have allowed them to adapt more easily than non-birdlike dinosaurs. Most birds are also able to fly, possibly making it easier for them to escape dangerous situations and find food in new places.

The evolution and variety of beaks probably helped birds survive as well. Beaks enable birds to kill prey, make nests, feed their young, and more, helping them thrive in different habitats. Bird species have different beaks, depending on their needs. A hummingbird has a long, narrow beak for drinking **nectar** from flowers. Warblers' pointy beaks enable them to pluck insects from logs and leaves. Eagles are meat-eating birds that use their sharp, hooked beaks to tear up their prey.

Today, there are at least 50 billion wild birds in the world, divided into more than 10,000 different species. About 1.6 billion of these birds are house sparrows, while European starlings number about 1.3 billion. Other species are very rare. For example, the giant ibis is critically **endangered** and numbers fewer than 200.

Woodpeckers have strong, pointy beaks to drill holes into wood.

To help it catch fish, a pelican has a long beak with a pouch.

A toucan's beak has jagged edges for skinning and cutting fruit. Its bright color also attracts mates.

Are all bird feathers the same?

All feathers are made of keratin, the same **protein** that makes up human nails and hair. However, there are seven different types of feathers: tail, wing, contour, semiplume, filoplume, bristle, and down. Stiff tail feathers help birds steer and stay balanced while flying. Wing feathers help birds lift their bodies into the air and power flight. Contour feathers cover a bird's body, while semiplume feathers lie between the contour feathers. Semiplume and down feathers, underneath, help with **insulation**. Bristle feathers, around the eyes and beak, and small filoplume feathers, on the body or wings, help birds sense temperature, wind speed, and more.

Wing feathers, also called flight feathers or remiges, are connected to the wing bones.

Most birds have six pairs of tail feathers, but birds can have as many as 16 pairs.

Do all birds fly?

Most birds fly, but there are exceptions, such as penguins and ostriches. These birds probably descended from flying birds, but evolved to adapt to their environments. Perhaps there were no predators that they needed to escape by flying. Instead, some birds developed excellent running or swimming abilities.

Penguins, for example, walk awkwardly on land and cannot fly, but they are excellent swimmers. Some species, such as the gentoo, can swim at speeds of up to 22 miles per hour (35 km/h). Penguins' wings have evolved into powerful flippers that they use to propel themselves through the water.

DID YOU KNOW?

The peregrine falcon is the world's fastest flier. When diving for its food, it can reach a speed of more than 186 miles per hour (300 km/h). In comparison, most birds fly at around 25 miles per hour (40 km/h).

A peregrine falcon's body is ***streamlined****, meaning it has a shape that encounters almost no* ***air resistance*** *as it dives toward its prey.*

An ostrich can run as fast as 43 miles per hour (69 km/h).

Why is a flamingo pink?

When flamingo chicks hatch, they have white-gray feathers. It is only as they grow older that they turn pink. Once a chick starts eating on its own, it will feast on algae, brine fly **larvae**, and brine shrimp. These wetland organisms, or living things, have high amounts of beta-carotene, a chemical with a red-orange pigment. In flamingos, the pigment turns their bodies different shades of red, orange, or pink. The shade a flamingo turns depends on the environment and food sources where it lives. Flamingos in the Caribbean tend to be brighter pink.

*Other animals get their color from foods too. Northern cardinals, which are bright red, also eat food high in **carotenoids**.*

Flamingos eat with their heads upside down. They take in water and food, shake their beak back and forth, and use their tongue to force water out and filter the food.

Why do birds sing?

Songbirds sing to protect their territories and to attract mates. Do songbirds also sing for fun? It is difficult to know, but studies show that singing releases dopamine—a chemical related to feeling happy—in a bird's brain. This chemical also encourages baby songbirds to learn to sing.

About half of all birds are songbirds. Other birds still make sounds to communicate, but songbirds perform an arrangement of more complex sounds. Birds have a voice box called a syrinx that is made up of two sides. Songbirds can control each side. Some birds can even make two sounds at once.

DID YOU KNOW?

Songbirds can be male or female. Scientists used to think that only male birds could sing. Recent studies, however, found that females sing in about 64% of all songbird species, such as barn swallows (below). Sometimes, males and females sing together.

Male and female birds can often be distinguished by their differing songs.

The North American brown thrasher sings at least 1,100 different songs. It includes imitations of other birds' songs in its performances.

How long can a plant live

Plants have life cycles ranging from a few weeks to many years. Scientists have identified 390,000 **vascular** plant species, including flowering plants, and almost 20,000 non-vascular plant species, such as mosses.

Plants change water and **carbon dioxide** into oxygen and sugars during photosynthesis—a process essential for all other life on Earth. Land plants first evolved from green algae about 500 million years ago and quickly spread to create habitats for land animals. To survive on land, plants developed vascular tissue that could take in water and **nutrients**. Plants then evolved to have seeds, and later, produce flowers. Flowering plants make up most of the plant species on Earth.

Paleobotanists study fossilized plants. In 2015, paleobotanists discovered the remains of *Montsechia vidalii* in Spain. The flowering plant is about 130 million years old and would have grown in shallow fresh water. Trees are some of Earth's oldest plants. About 25 different species, including cypress, baobab, and ginkgo trees, can live for 1,000 years or longer. Baobabs grow naturally in dry areas in Africa and Australia. They are the only flowering plants that can live over 1,000 years.

The baobab tree is known as the tree of life. It provides food, water, and shelter for animals and humans.

Scientists discover around 2,000 new plants each year!

Montsechia vidalii is believed to be one of the first flowering plants.

The longest living plant of all is the bristlecone pine, which lives for thousands of years. These trees are found in mountainous regions of the southwestern United States. The world's oldest plant is a 5,060-year-old bristlecone pine in the White Mountains of California.

Scientists date trees by studying the **growth rings** in a tree's trunk. Most tree species produce a new growth ring each year. However, the bristlecone pine's inner rings are very tight together and wear away over time. This makes it difficult to determine an accurate date.

The bristlecone pine may be the oldest individual plant, but clonal colonies can be thousands of years older. A clonal colony is a group of identical organisms that grow from one original organism. Many plants grow in clonal colonies, but the biggest and oldest is Pando.

The bristlecone pine is a slow-growing tree. Over many years, harsh winds and cold temperatures twist the dense trunks and branches.

Since the trees in Pando all share the same root system, they change color and lose leaves at the same time.

Pando is a clonal colony of quaking aspen trees in Utah's Fishlake National Forest. It covers about 106 acres (43 ha). Pando means "I spread" in Latin. It looks like a forest of over 40,000 individual trees, but they are all connected to a single root system that expanded from a single seed. At almost 13 million pounds (5.9 million kg), it is the largest and most **dense** organism ever found. Researchers estimate that the original seed and root system is up to 12,000 years old.

What is a mother tree?

There are many individual trees in a forest. But beneath the ground, they are connected through their root systems. Scientific studies show that trees have developed amazing ways to communicate and help each other through these underground connections.

This underground system is known as the mycorrhizal network. It is made up of fungi that grow around and inside the tree roots. The fungi grow tiny threads called **mycelium**, which spread out into a large web and connect to more roots to build a network of connected plants.

Mother trees, or hub trees, are an essential part of this network. They are old, tall trees with deep roots and many fungal connectors. They use the network to share water and nutrients with smaller, younger trees that have less exposure to sunlight. The network also allows mother trees to receive distress signals from struggling trees and send them extra nutrients.

Scientists are also listening to trees. Using special instruments, they can hear crackling in the roots. They have also detected electrical and scent signals. Scientists believe the trees are sending distress messages to each other. For example, if an animal is eating a tree's leaves, it sends chemical signals that warn other trees to defend themselves, such as using chemicals to make their leaves taste bad.

The nickname for this tree network is the "wood-wide web."

Through the mycelium, the fungi joins with tree roots.
Mushrooms are part of the underground fungal organism.
Young, small trees, called saplings, depend on mother trees for survival.
A mother tree might be connected to hundreds of other trees in a forest.

Are there meat-eating plants?

Most plants produce their own food through photosynthesis. A few types of plants, however, add to their diet by eating insects. Usually, they do this because there are not enough nutrients in the soil. There are more than 600 known types of carnivorous, or meat-eating, plants, including the Venus flytrap, the pitcher plant, and the Australian sundew.

Carnivorous plants trap prey in different ways. Insects fall into pitcher plants and cannot get out. Australian sundews and some bladderworts trap prey with sticky leaves. Venus flytraps snap their leaves shut onto unsuspecting prey. The carnivorous plants produce **enzymes** that dissolve the prey so it can be used for food.

Giant montane pitcher plant

Venus flytrap

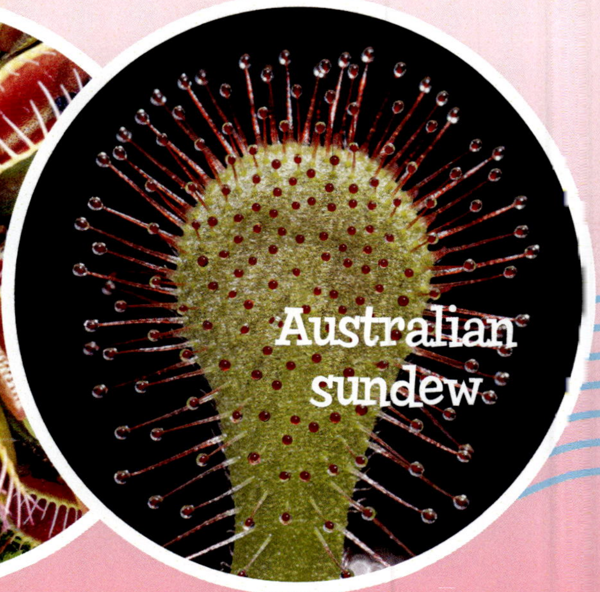

Australian sundew

What kinds of plants live in the ocean?

Marine plants thrive in salty oceans. They rely on sunlight, so they grow in shallow waters. Algae look like plants, but they are classified as protists. There are many different types of algae, including giant kelp and species of seaweed.

Seagrasses are flowering plants that grow in coastal waters around the world, except Antarctica. These plants evolved about 100 million years ago, and there are now about 72 different species. Seagrasses are crucial to ocean ecosystems. They provide animals with food and habitats, clean the water, and produce huge amounts of oxygen.

Seagrass, Shark Bay

Giant kelp is a type of brown algae that grows to 100 feet (30.5 m). Huge kelp forests provide food and shelter for many ocean creatures.

DID YOU KNOW?

Some seagrass meadows are large enough to be seen from space! A clonal colony of seagrass in Shark Bay, Australia, covers about 77 square miles (200 sq. km), an area larger than Washington, D.C. The colony is about 4,500 years old and expanded from a single plant. It is the world's largest living plant.

How many insects are there in the world?

There are more kinds of insects than any other organism on Earth. Scientists estimate that there are at least 1 million, but that there could be as many as 10 million. The total number of insects could be at least 10 quintillion (10,000,000,000,000,000,000)! Insects have adapted to live in habitats all over the world and are essential to those ecosystems.

Insects are arthropods, an animal group that includes crabs, spiders, and millipedes. Arthropods have an **exoskeleton**, a body with three parts, and at least three pairs of legs. Insects are divided into 29 different groups, including beetles, moths and butterflies, and ants.

Insects may seem annoying and some even cause disease, but the world needs them. Insects are the main food for many animals. Many people eat insects too. Insects **pollinate** plants, control other pests that damage food crops, and break down organic waste for healthy soil. Human activities such as habitat destruction and pollution have threatened insect survival, however.

The white witch moth is one of the world's largest insects, with a wingspan up to 14 inches (36 cm).

The hercules beetle is the longest beetle at 7 inches (18 cm).

The Dinoponera is the world's largest ant, at more than 1.2 inches (3 cm) long.

The world's smallest insect is the fairyfly—a tiny wasp measuring just 0.005 inches (0.127 mm) long.

Could spiders get caught in their own webs?

Spiders are arthropods, like insects, but they form a completely separate group called arachnids. Spiders have two-part bodies, eight legs, and six or eight eyes.

Most spiders do not make webs at all, but all spiders produce silk. Scientists have a few ideas about how spiders avoid getting tangled. Some of the silk is not sticky, so spiders may avoid these threads on a web. Also, web-spinning spiders have an extra set of claws on their feet. They can use these to grab and move around the web.

Spiders spin silk using glands that produce a watery, gel-like substance. It becomes silk as it shoots out of organs called spinnerets. But not all silk is the same. The orb-web spider produces at least four different kinds of silk, making it stronger or stickier, depending on its needs. Webs often catch prey but can also attract a mate, or protect a spider's egg sac.

DID YOU KNOW?

Some spiders can fly! They do not have wings, but can launch themselves through the air in a maneuver called ballooning. They spin out some silk and use wind and Earth's electric field to propel themselves high in the air and as far as 1,000 miles (1,609 km). Ballooning is used by spiders to get from place to place.

There are about 46,700 known spider species, and they are all predators.

The golden silk orb-weaver is known for the golden color of its silk.

Funnel-web spiders live in moist, sheltered places such as under rocks. They burrow into the ground and weave their web around the entrance.

Most spiders, like the tarantula, have venom glands, but their bites are rarely dangerous to humans.

Do insects have favorite colors?

It seems that insects might prefer some colors more than others, but scientists are still investigating. Butterflies seem to like red, pink, yellow, and orange flowers. This may be because they have learned that these flowers have the best nectar. Honeybees seem to prefer purples and blues.

How insects see color varies from species to species. Many insects see only some colors. Most have two types of color pigments in their eyes. Most, except butterflies, cannot see lower wavelengths like red. However, they may be able to see reddish colors like orange and yellow.

Most insects have compound eyes, made up of hundreds or thousands of tiny parts called ommatidia. A dragonfly's eye has about 28,000.

Unlike humans, many insects can see ultraviolet light. This picture shows how a flower looks under ultraviolet light.

Monarchs fly during the day and rest at night. They choose to rest in trees with thick canopies, where the temperature will be not too hot or cold.

Do monarch butterflies get tired when they migrate?

Each fall, eastern monarch butterflies migrate from the northeastern United States and southeastern Canada to central Mexico. For some, the journey is up to 3,000 miles (4,828 km). Monarchs can fly as far as 100 miles (161 km) in a day, but they do need to rest and feed on nectar along the way.

When they arrive in Mexico, monarchs hibernate in oyamel fir trees. In early spring, the monarchs start heading north. They search for milkweed plants to lay their eggs. This is the only plant monarch larvae will eat. It is poisonous to some animals and may help the monarch protect itself.

Researchers are testing an idea that monarchs benefit from having larger white spots on their wings. These spots may change the air flow and temperature and help the butterflies to survive the long, hot journey.

Can insects live in the Arctic?

Many insects cannot survive the cold temperatures in the Arctic, which can reach as low as –50 to –65 degrees Fahrenheit (–46 to –54 degrees Celsius) in some regions. Despite this, some insects do live in the Arctic. Some migrate in winter, but those that do stay all year have adapted to the frigid weather. Some find shelter and go into a kind of hibernation called diapause. Others freeze solid, but keep the ice from harming important organs by first removing most of the water from their bodies.

The Arctic woolly bear caterpillar, for example, produces proteins that act as an antifreeze in its body. When it finally becomes a moth in early summer, it mates and lays eggs. The adult moth dies soon afterwards and the eggs hatch quickly into caterpillars.

The red flat bark beetle survives by producing antifreeze proteins and ridding its body of water. At the same time, it produces a substance called **glycerol**, which cools it down. This combination of chemicals keeps the beetle in a glass-like state called vitrification.

The Arctic woolly bear caterpillar is only active and feeding in the short summers. It takes seven years or longer to become a moth.

DID YOU KNOW?

Bees can live in the Arctic! The Arctic bee has some amazing adaptations that allow it to survive in extremely cold Arctic temperatures. On sunny days, it sits inside cone-shaped flowers, which act as mini magnifying glasses to intensify the Sun's rays. Arctic bees also have large flight muscles, which they shiver to increase their body temperature.

Arctic bees have more hair than most other bee species. Their thick coat helps them stay warm.

The red flat bark beetle can survive temperatures as low as -112 °F (-80 °C).

Glossary

air resistance (air ri-ZIS-tuhns): The force of air moving against an object

algae (AL-jee): A group of plant-like organisms that grow in or near water

bioluminescence (bahy-oh-loo-muh-NES-uhns): Light produced by organisms through chemical reactions

carbon dioxide (KAHR-buhn dahy-OK-sahyd): A colorless gas used by plants in photosynthesis

carotenoids (kuh-ROT-n-oids): Natural red, orange, and yellow pigments

dense (dens): Having a lot of matter packed closely together

endangered (en-DEYN-jerd): At risk of becoming extinct

enzymes (EN-zahyms): Proteins in living things that help speed up chemical reactions

evolved (ih-VOLVD): Of an animal, gradually changed over time to better suit its environment

exoskeleton (ek-soh-SKEL-i-tn): A hard outer layer that protects the body of an organism

fossil (FOS-uhl): Preserved remains of an organism

fungi (FUHN-gahy): Organisms, such as mushrooms, that are neither plants or animals

glands (glands): Organs that release substances for certain purposes in the body

glycerol (GLIS-uh-rawl): A liquid produced by insects to make their body fluids too viscous, or thick, to freeze

growth rings (grohth rings): In tree trunks, circular rings that form each year, showing the tree's age

insulation (in-suh-LEY-shuhn): Covering something to protect it or stop heat or sound from escaping

Jurassic period (joo-RAS-ik PEER-ee-uhd): The time in Earth's history in which dinosaurs thrived

larvae (LAHR-vey): Immature, often wormlike, form of an animal or insect

migrate (MAHY-greyt): To move from one area to another with the seasons

mycelium (mahy-SEE-lee-uhm): Root-like structure of a fungus

nanocrystals (NAN-oh-KRIS-tls): Very tiny crystal particles only visible with an electron microscope

nectar (NEK-ter): Sugary liquid produced by plants

neurons (NOOR-ons): Nerve cells; basic units of the nervous system

nutrients (NOO-tree-uhnts): Substances used by living things to live and grow

pigments (PIG-muhnts): Substances that give something color

pollinate (POL-uh-neyt): To move pollen from one plant to another so new plant seeds can be produced

protein (PROH-teen): Substances found in living organisms that do important work in the body's cells

protists (PROH-tists): A group of mostly single-celled organisms that are not bacteria, fungi, animals, or plants

swim bladder (swim BLAD-er): An organ that keeps fish from sinking

tadpoles (TAD-pohls): Amphibian larvae that have tails and breathe through gills

theropod (THEER-uh-pod): A meat-eating dinosaur that walks on its back legs

vascular (VAS-kyuh-ler): In plants, having a system of tissues that move water and nutrients

wavelengths (WEYV-lengkths): The distance of a sound or light wave

Index

Comprehension Questions

1. **True or False:** Less than 100 species of plants are carnivorous.
2. **__________ trees are the oldest flowering plants on Earth.**
   a) Baobab
   b) Cypress
   c) Ginkgo
3. **A dragonfly's eye has about __________ ommatidia.**
   a) 21,000
   b) 35,000
   c) 28,000
4. **True or False:** Alligators have salt glands that allow them to live in both salt water and fresh water.
5. **True or False:** Elephants have the largest brains of all land animals.
6. **The __________ survives Arctic winters by producing glycerol in its body.**
   a) Arctic woolly bear caterpillar
   b) red flat bark beetle
   c) Arctic bee
7. **A __________ has a beak with jagged edges for skinning and cutting fruit.**
   a) woodpecker
   b) toucan
   c) pelican
8. **True or False:** In Latin, Pando means "I spread."

1. False, 2. a, 3. c, 4. False, 5. True, 6. b, 7. b, 8. True

About the Author

Cynthia O'Brien's curiosity about the world led her to be a writer for young readers. She has written over 80 books covering a wide range of subjects.

Crabtree Publishing

crabtreebooks.com 800-387-7650

 In Canada: We acknowledge the financial support of the Government of Canada through the Canada Book Fund for our publishing activities.

Hardcover 978-1-0398-8093-1
Paperback 978-1-0398-8454-0
Ebook (pdf) 978-1-0398-8213-3
Epub 978-1-0398-8333-8

Printed in the U.S.A./CP2025

Published in Canada
Crabtree Publishing
616 Welland Avenue
St. Catharines, Ontario
L2M 5V6

Published in the United States
Crabtree Publishing
347 Fifth Avenue
Suite 1402-145
New York, NY 10016

Written by: Cynthia O'Brien
Series development: Janine Deschenes
Editorial coordinator: Kathy Middleton
Editor: Janine Deschenes
Proofreader: Melissa Boyce
Art direction and layout: Katherine Kantor and Margaret Salter
Photo research: Katherine Kantor
Production manager: Candice Campbell

Library and Archives Canada Cataloguing in Publication
Available at the Library and Archives Canada

Library of Congress Cataloging-in-Publication Data
Available at the Library of Congress

Images:
Flickr: USAID Biodiversity & Forestry, WCS: p. 24 (bird)
Wikimedia:
Kuba Bozanowski from Warsaw, Poland: p. 6–7
Free Public Domain Illustrations by rawpixel: p. 10 (fish)
Daderot: p. 13 (skeleton)
Kumiko: p. 22 (bottom)
Luis Fernández García: p. 31 (bottom right)
J. Zapell: p. 33
Huber J.T.: p. 39 (bottom right inset)
Mike Beauregard from Nunavut, Canada: p. 44 (caterpillar)
All other images by Shutterstock